THE POETRY OF INDIUM

The Poetry of Indium

Walter the Educator

SKB

Silent King Books a WhichHead Imprint

Copyright © 2023 by Walter the Educator

All rights reserved. No part of this book may be reproduced in any manner whatsoever without written permission except in the case of brief quotations embodied in critical articles and reviews.

First Printing, 2023

Disclaimer
This book is a literary work; poems are not about specific persons, locations, situations, and/or circumstances unless mentioned in a historical context. This book is for entertainment and informational purposes only. The author and publisher offer this information without warranties expressed or implied. No matter the grounds, neither the author nor the publisher will be accountable for any losses, injuries, or other damages caused by the reader's use of this book. The use of this book acknowledges an understanding and acceptance of this disclaimer.

"Earning a degree in chemistry changed my life!"
– Walter the Educator

dedicated to all the chemistry lovers, like myself, across the world

CONTENTS

Dedication v

Why I Created This Book? 1

One - Indium, The Celestial Metal 2

Two - Higher And Higher 4

Three - Symphony Of Progress 6

Four - The Enigma 8

Five - Possibilities Sway 10

Six - Here To Stay 12

Seven - Rejoice 14

Eight - Serene 16

Nine - Free 18

Ten - Mold 20

Eleven - Time Flies By 22

Twelve - Destiny 24

Thirteen - Innovation's Ally	. . . .	26
Fourteen - Ode To Prevail	. . . .	28
Fifteen - Hand In Hand	. . . .	30
Sixteen - For Me And For You	. . . .	32
Seventeen - Sustains Our Planet	. . . . . .	34
Eighteen - World That's Anew	. . . .	36
Nineteen - Purpose Divine	. . . .	38
Twenty - Cosmic Chase	. . . .	40
Twenty-One - Eternal Fable	. . . .	42
Twenty-Two - Conductor Of Change	. . . . .	44
Twenty-Three - Bound To Be Bright	. . . .	46
Twenty-Four - Illuminates Us All	. . . . .	48
Twenty-Five - Burning Bright	. . . .	50
Twenty-Six - Rise And Stand	. . . .	52
Twenty-Seven - Beautiful Earth	. . . . . .	54
Twenty-Eight - Symbol Of Hope	. . . . . .	56
Twenty-Nine - Across The Land	. . . . . .	58
Thirty - We Shall Seek	. . . .	60
Thirty-One - Radiant Light	. . . .	62
Thirty-Two - Conductive Charm	. . . . . .	64

Thirty-Three - Brilliance 66

Thirty-Four - Indium's Significance 68

Thirty-Five - Awed 70

About The Author 72

WHY I CREATED THIS BOOK?

Creating a poetry book about the chemical element Indium can be an intriguing and innovative idea. Indium, being a lesser-known element, offers a unique opportunity to explore its properties and symbolism through poetry. By delving into its atomic structure, historical significance, or industrial applications, you can craft verses that highlight its distinct characteristics. You can also use Indium as a metaphor, drawing parallels between its properties and various aspects of life, emotions, or relationships. Such a poetry book can be an artistic exploration of the intersection between science and creativity, appealing to both science enthusiasts and poetry lovers.

ONE

INDIUM, THE CELESTIAL METAL

In the depths of Earth's embrace, a hidden treasure lies,
A metal born of magic, where secrets crystallize.
Indium, the alchemist's gift, with lustrous silver sheen,
A tale of wonder and beauty, a marvel to be seen.

Born in molten fires, deep within the Earth's core,
Indium emerges, a radiant element to adore.
Its atoms dance in harmony, in a lattice so fine,
A symphony of electrons, intertwining divine.

With properties unique, it bends and it reflects,
A mirror to the world, where light fondly connects.
From liquid crystal displays to solar cells so bright,
Indium's versatility brings forth a dazzling light.

A conductor of electricity, it powers our modern

age,
Transmitting energy through circuits, an electrical sage.
In touchscreens it thrives, with a gentle tap it responds,
Creating a world of wonders, where technology absconds.

But beyond its scientific wonders, Indium holds a mystery,
A connection to the cosmos, a cosmic tapestry.
For within its very core, the stars themselves reside,
As supernovae's remnants, in Indium they confide.

So let us celebrate this element, this treasure from afar,
Indium, the celestial metal, a shining cosmic star.
In laboratories and beyond, its wonders we explore,
As we unravel the secrets of this element we adore.

TWO

HIGHER AND HIGHER

In the depths of nature's secret forge,
A hidden treasure waits to emerge,
Born of magic, Indium takes its place,
With grace and charm, it adorns our space.
    A metal rare, with lustrous sheen,
Indium, a marvel, seldom seen,
Its properties diverse, its virtues vast,
A versatile element, unsurpassed.
    Indium, the conductor of our dreams,
In screens and circuits, its brilliance gleams,
A vital link in technology's chain,
Enabling progress, removing the strain.
    Yet within its core, lies a cosmic tale,
Remnants of supernovae, a celestial trail,

From distant stars, it found its way,
To Earth's embrace, where it shall stay.
    Oh, Indium, celestial metal bright,
A bridge between worlds, day and night,
We marvel at your infinite worth,
And explore your secrets, from birth to rebirth.
    So let us cherish this gift from above,
Indium, the element we truly love,
A testament to our boundless desire,
To conquer the unknown, to reach higher and higher.

THREE

SYMPHONY OF PROGRESS

Indium, a hidden treasure born of magic,
A lustrous silver sheen, elusive and enigmatic.
In the realm of elements, you hold a special place,
With versatile properties, a marvel to embrace.

Conductor of electricity, you bring energy to life,
Linking screens and circuits, with precision and strife.
In the tapestry of technology, you weave a vital thread,
Guiding signals, transmitting power, as we forge ahead.

Within your core, remnants of supernovae reside,
Cosmic stardust, a celestial journey we cannot hide.
From the depths of the universe, you found your way,
A cosmic creation, in our world you stay.

Indium, rare and precious, a gem we hold dear,

A conductor of dreams, bridging realms far and near.
You connect our desires, our hopes, and our dreams,
A catalyst for innovation, you reign supreme.

Oh, Indium, we admire your secrets untold,
In labs and workshops, your mysteries unfold.
As we delve into your essence, we unravel your truth,
With every discovery, we're in eternal youth.

So let us celebrate your presence, oh divine Indium,
A marvel of science, a key to our rhythm.
In screens and circuits, you silently play,
A symphony of progress, guiding our way.

FOUR

THE ENIGMA

In the depths of the Earth, a treasure lies,
A metal rare, with secrets it implies.
Indium, the name that echoes through time,
A lustrous silver sheen, so sublime.

Conductor of currents, electricity's friend,
In technology's realm, its wonders extend.
From touchscreens to solar cells, it plays its part,
Indium's versatility, a work of art.

Born in cosmic explosions, a celestial dance,
Supernovae remnants, within its core enhance.
A gift from the stars, a cosmic embrace,
Indium's origins, an awe-inspiring trace.

Now, a new tale, a different refrain,
Indium, enigmatic, its mysteries remain.

A thread in the tapestry of progress and dreams,
Connecting desires, igniting innovation it seems.
A catalyst it becomes, a spark in the dark,
Guiding our footsteps, leaving a lasting mark.
In the realm of technology, it finds its place,
Indium's presence, a testament to grace.
Oh, Indium, we celebrate your silent might,
Your shimmering presence, a guiding light.
Invisible conductor, a silent force,
You shape our world, with your endless source.
So, let us cherish this element rare,
Indium, the enigma, beyond compare.
For in your hidden depths, a universe thrives,
Indium, the element, where progress arrives.

FIVE

POSSIBILITIES SWAY

In the depths of the cosmos, where stars are born,
Lies a shimmering element, Indium, adorned.
A dance of electrons, a symphony of light,
This precious metal, a cosmic delight.

With a silvery sheen, it catches the eye,
A whisper of beauty, as the moonlight sighs.
Indium, the alchemist's dream, so rare,
Transforming the ordinary with a touch so fair.

In the realm of technology, it finds its place,
A conductor of dreams, a celestial embrace.
From screens that glow with colors bold,
To soldering bonds that never grow old.

Indium, the connector, bridging the gap,
Between circuits and chips, a technological map.

A catalyst for progress, innovation's guide,
It fuels the fires of change, spreading far and wide.
    From solar panels that harness the sun's embrace,
To touchscreens that respond with gentle grace,
Indium, the wizard, creating wonders untold,
Invisible threads of magic, forever to behold.
    So let us celebrate this gift from above,
Indium, the element, that inspires love.
For in its versatile nature, we find our way,
To a future where dreams and possibilities sway.

SIX

HERE TO STAY

In the realm of elements, a star does reside,
Indium, the name that makes our hearts wide,
A metal of wonder, a shimmering light,
In technology's realm, it shines ever so bright.
    From liquid crystal displays to touchscreens of grace,
Indium's versatility, we eagerly embrace,
Conducting electricity with prowess untold,
Its presence, a marvel, a story to be told.
    In soldering's embrace, it binds with finesse,
Connecting the pieces, a delicate caress,
With its low melting point, it dances with ease,
Uniting the fragments, creating symphonies.
    Indium, the magician, in semiconductors it dwells,
Enhancing the signals, where innovation swells,

From transistors to diodes, it weaves its magic,
Guiding the currents, a technological fabric.

Oh, Indium, how you soar, like a cosmic star,
Born in supernovas, a journey from afar,
From the depths of the universe, you came to be,
A celestial gift, connecting you and me.

So let us celebrate, this element so grand,
Indium, the catalyst, that sparks progress's hand,
In the realm of science, it paves the way,
A shining beacon, forever here to stay.

SEVEN

REJOICE

In the realm of science, a secret lies,
A hidden gem that dazzles our eyes.
Indium, the conductor of dreams,
Unveils the path to endless schemes.

A bridge between worlds, it does create,
Connecting the dots of our technological fate.
In screens it dances, pixels aglow,
In circuits, it hums, a symphony's flow.

From supernovae, it was born,
Cosmic stardust, a tale to adorn.
Mysterious origins, a celestial birth,
A whisper of stardust, sprinkled on Earth.

Oh, Indium, you catalyst of progress,
Innovation's muse, you graciously possess.

Semiconductors bow before your might,
Guiding us towards an illuminated sight.
   With soldering touch, you bind and unite,
Elements disparate, in harmonious light.
A silent force, shaping our world,
Unleashing wonders, yet to be unfurled.
   Indium, the creator of endless arrays,
Of screens that mesmerize and circuits that blaze.
In your versatility, we find our stride,
A symphony of progress, forever in your guide.
   So, let us celebrate this element grand,
Indium, the conductor of our technological band.
With gratitude and awe, we raise our voice,
To honor your essence, in you we rejoice.

EIGHT

SERENE

Indium, conductor of dreams,
In your metallic embrace, progress gleams.
A catalyst for innovation and thought,
You guide us where the imagination is sought.
 Oh, Indium, versatile and true,
Connecting the world, shaping it anew.
Through your conductive touch, technology thrives,
As we build a future where dreams come alive.
 In the realm of transistors and screens,
You illuminate the path to digital dreams.
From laptops to smartphones, you play your part,
Indium, the magician, igniting the spark.
 With solder in hand, you work your charm,
Binding components, keeping them warm.

A guiding force in the realm of creation,
Indium, the magician of scientific revelation.
    And as we delve deeper into the unknown,
You stand as a symbol, never alone.
Uniting the elements, forging a bond,
Indium, the silent force, forever beyond.
    Oh, Indium, you shape our world with grace,
Unleashing wonders, leaving no trace.
A conductor of harmony, a force unseen,
Indium, the element of dreams, serene.

NINE

FREE

In the realm of dreams and innovation,
There lies a metal, Indium by name.
A conductor of thoughts, a catalyst of creation,
In its presence, brilliance will surely proclaim.

Indium, the guide, the beacon of light,
Leading us forward, through the darkest night.
A bridge between worlds, a spark that ignites,
The flames of progress, burning ever so bright.

With its touch, connections are made,
From circuitry's web, a symphony played.
Binding the wires, in harmony they sway,
Indium, the magician, casting its display.

Versatile it is, in its forms and states,
Malleable and fluid, it resonates.

In screens it shimmers, a vibrant display,
A silent force shaping our world day by day.
　　Indium, the alchemist, transforming the mundane,
Into wonders untold, a future unrestrained.
From technology's embrace, it never strays,
Guiding us towards possibilities that never fade.
　　So let us honor this element divine,
Indium, the conductor, weaving the design.
In its presence, dreams become reality,
Indium, the silent force that sets us free.

TEN

MOLD

In the realm of technology's embrace,
A magician dances with elegant grace.
Indium, the catalyst of progress profound,
A shimmering jewel, forever renowned.

    With semiconductors, it weaves a spell,
Connecting dreams, where innovation dwells.
A conductor of harmony, it unites,
Elements of wonder, in circuitry's bright lights.

    Indium, the alchemist of our age,
Turning dreams into reality, page by page.
From smartphones to tablets, screens so vivid,
Its touch brings life, as if by magic it's guided.

    A metal so rare, yet abundantly found,
Indium shapes our world, without a sound.

With solder's embrace, it binds and connects,
A symphony of creation, where dreams intersect.
    In laboratories, its secrets unfold,
As scientists marvel, their wonders behold.
The mysteries it holds, still yet to be known,
Indium's allure, forever it's shown.
    Let us celebrate this element divine,
A guiding force, where brilliance will shine.
Indium, the conductor of dreams untold,
Innovation's partner, forever it'll mold.

ELEVEN

TIME FLIES BY

In the realm of progress and innovation,
A silent conductor, Indium, takes its station.
A metal with grace, its presence discreet,
Indium's touch makes wonders complete.

In screens and circuits, it finds its place,
A pixel's dance, a vibrant embrace.
With every touch, its brilliance shines,
Indium's magic, a symphony of lines.

From smartphones to laptops, it lends its might,
Guiding us through the digital night.
A conductor of dreams, it paves the way,
Indium's whispers, the songs we play.

In laboratories, it sparks new frontiers,
Scientists rejoice, as it disappears.

Indium's secrets, a puzzle to solve,
Unveiling mysteries, as they evolve.
    In solar panels, it captures the sun,
Harnessing energy, as day is done.
Indium's power, a gift from above,
Nurturing our planet with boundless love.
    A conductor of progress, it shapes our world,
Indium's symphony, forever unfurled.
From the depths of the earth, it rises high,
Guiding us forward, as time flies by.

TWELVE

DESTINY

In the realm of progress, where dreams take flight,
There lies a noble element, shining so bright.
Indium, the conductor of innovation's song,
Guiding us forward, where we belong.

Within its atomic embrace, wonders unfold,
A metal with power, yet modest and bold.
In the realm of technology, it finds its place,
Binding us all in a digital embrace.

Indium, the alchemist of screens and displays,
A magician of light, in a myriad of ways.
From LCD to OLED, its magic unfurls,
Transforming our world, with colors that swirl.

In the realm of semiconductors, it holds the key,
Unlocking the secrets of connectivity.

A bridge between worlds, it paves the way,
Uniting electrons, in a harmonious display.
    Indium, the guardian of solar panels' might,
Harnessing the sun's energy, day and night.
A beacon of hope, in a world that craves,
Sustainable power, for the future it saves.
    In every field, Indium weaves its spell,
A silent force, in the tales we tell.
Connecting and binding, with a touch so fine,
Unleashing possibilities, that forever shine.
    Oh, Indium, you shape our reality,
A catalyst for progress, for all to see.
With each new discovery, you set us free,
Indium, the element, that guides our destiny.

THIRTEEN

INNOVATION'S ALLY

In the realm of technology, you shine bright,
Indium, the element of innovation's might.
With atomic number forty-nine you stand,
A conductor of progress across the land.

   Your silvery hue, so sleek and refined,
Reflects the wonders of mankind.
A metal with a heart so pure,
Indium, you're the catalyst we adore.

   From touchscreens to transistors, you play a vital role,
Guiding us through the depths of the digital soul.
Your malleable nature, a sculptor's dream,
Transforming the world with a gleaming gleam.

   Indium, you connect and unite,
Joining elements with a bond so tight.

In solder, you bind components with grace,
Fostering communication in every place.
   Oh Indium, you're a star in the sky,
In solar panels, you help the sun's energy fly.
Harnessing the power of the radiant light,
You pave the way for a future so bright.
   So let us celebrate your atomic grace,
Indium, the metal that leaves no trace.
A symbol of progress, a conductor of change,
Innovation's ally, forever you'll range.

FOURTEEN

ODE TO PREVAIL

In the realm of elements, a shining star,
Indium, the conductor, has traveled far.
A metal with a purpose, so pure and bright,
Igniting dreams and illuminating the night.

Technology's ally, Indium takes its place,
Uniting circuits, weaving a web in space.
From smartphones to screens, its touch is felt,
A symphony of progress, a melody to be dealt.

Indium, the alchemist, with transformative might,
Bridging the gaps, turning darkness into light.
With soldering hands, it joins the tiniest parts,
Creating connections, nurturing endless arts.

In solar panels, its magic is seen,
Harnessing the sun's rays, a renewable dream.

A champion of sustainability, it plays its role,
Shaping a future where harmony takes its toll.

    Indium, the catalyst, ignites innovation's fire,
Fueling the dreams of those who aspire.
With every breakthrough, a step closer we tread,
A world united, where possibilities spread.

    Oh Indium, the guide, with wisdom profound,
Leading us forward, where new horizons are found.
With grace and purpose, it paves the way,
A beacon of hope, lighting our path each day.

    So let us celebrate this wondrous element,
Indium, the connector, so vibrant and resplendent.
In science, in life, it weaves its tale,
A testament to its power, an ode to prevail.

FIFTEEN

HAND IN HAND

In the realm of elements, a marvel we find,
Indium, the conductor, so gentle and kind.
With atomic number forty-nine, it shines,
In the periodic table, where harmony aligns.

A treasure, rare and precious, it bestows,
A conductor of electricity, it proudly shows.
From the finest screens to circuits profound,
Indium leads the way, a conductor renowned.

In solder and alloys, it lends a hand,
Binding metals together, a magical band.
With grace and finesse, it forms a bond,
Uniting the elements, forever beyond.

In catalysts, it whispers, a secret it shares,
Accelerating reactions, removing life's cares.

A catalyst of change, it sparks the flame,
Igniting innovation, progress without shame.
　Indium, the guide, lighting up the way,
Leading us forward, to a brighter day.
In solar cells it basks, harnessing the light,
Sustaining our planet, with all its might.
　So let us celebrate this element divine,
Indium, the conductor, a gift so fine.
In technology's embrace, it takes its stand,
Connecting our world, hand in hand.

SIXTEEN

FOR ME AND FOR YOU

In the realm of technology, a hidden treasure lies,
A metal of wonders, that binds and unifies,
Indium, the element, with a gleam so divine,
A catalyst for progress, a conductor of time.

With soldering touch, it connects the unseen,
Linking circuits and pathways, a digital dream,
A bridge between realms, where innovation resides,
Indium, the conductor, where connectivity abides.

In screens it enchants, with colors so bold,
Displaying the world, in stories untold,
Touching our hearts, with each vibrant hue,
Indium, the artist, bringing wonders anew.

But beyond the screens, its powers extend,
To solar horizons, where energy transcends,

It harnesses the sun, with a gentle embrace,
Indium, the sustainer, in a renewable race.

In solar panels it dances, capturing the light,
Transforming it to power, a beacon so bright,
Guiding us forward, to a future so green,
Indium, the guardian, of a sustainable scene.

Indium, the element, so versatile and grand,
Binding worlds together, with a guiding hand,
From technology's touch, to the sun's golden ray,
Indium, the connector, lighting our way.

So let us embrace, this element divine,
In its unity lies, a future so fine,
Indium, the catalyst, for a world that's anew,
Shaping a brighter tomorrow, for me and for you.

SEVENTEEN

SUSTAINS OUR PLANET

In the realm of elements, Indium reigns,
A conductor of energy, it proudly sustains.
With atomic number forty-nine it's known,
A guardian of solar panels, it has shown.

Indium, the catalyst for progress and change,
In circuits and gadgets, it finds its range.
Connecting the dots, bridging the divide,
Innovation and technology, it does guide.

Malleable and soft, it adapts with ease,
Forming alloys, it unites with utmost peace.
A metal of grace, it bends and it molds,
Indium's presence, a story it unfolds.

A messenger of communication it becomes,
Binding us together, like sweet honeycomb.

In screens and displays, its brilliance displayed,
Indium, the connector, in every way portrayed.
 A champion of sustainability, it stands,
Embracing the future with eco-friendly plans.
Catalyzing green energy, a beacon so bright,
Indium guides us towards a world that's right.
 As the earth's resources we strive to preserve,
Indium's significance, we cannot observe.
A binder of metals, a catalyst of change,
Indium sustains our planet, it's truly a range.

EIGHTEEN

WORLD THAT'S ANEW

In the realm of atoms, a shimmering light,
Indium emerges, a radiant sight.
A conductor of progress, a catalyst true,
Binding elements, forging something new.

In solar panels, its magic unfolds,
Harnessing sunbeams, turning them gold.
Efficient and steadfast, it silently gleams,
Powering dreams with its solar streams.

Indium connects, it unites and combines,
Transforming the ordinary into the divine.
Like a bridge between worlds, it weaves its spell,
Creating harmony where chaos once dwelled.

With soldering grace, it joins the fray,
Fusing metals together, day after day.

In circuits and wires, it guides the flow,
Enabling innovation to flourish and grow.
   But Indium's true marvel lies in its heart,
Champion of sustainability, playing its part.
A guardian of Earth, a steward of green,
Lighting the path to a future pristine.
   So let's celebrate Indium, this magical star,
For its powers are endless, both near and far.
With its gentle touch and luminous hue,
It beckons us forward, to a world that's anew.

NINETEEN

PURPOSE DIVINE

In the realm of innovation, a catalyst shines,
Indium, the element, with powers so fine.
With atomic number forty-nine, it's true,
A metal rare, but its worth we pursue.

Indium, the connector, that binds metals tight,
Forging new paths, pushing boundaries with might.
In soldering, it glows with a silvery hue,
Uniting components, creating something new.

In technology's realm, where progress is key,
Indium paves the way, setting spirits free.
From touchscreens to solar cells, it plays a part,
Guiding us forward, igniting a spark.

Renewable energy, a future so bright,
Indium's sustenance fuels the world's fight.
In photovoltaic cells, it captures the sun,
Harnessing its power, till day is done.

Indium, a guide, in semiconductors grand,
Leading electrons, conducting the band.
With precision it navigates, without a hitch,
Lighting our path, like a lantern, rich.

So let us cherish Indium, this element rare,
A catalyst of progress, beyond compare.
Innovation's ally, it continues to shine,
Guiding us forward, with purpose divine.

TWENTY

COSMIC CHASE

Indium, the bridge that binds,
A connector, so refined,
In the realm of elements, you shine,
With a touch of magic, so divine.

In circuits and wires, you intertwine,
Guiding currents, a conductor so fine,
A catalyst for innovation, a spark of design,
Indium, the essence of technology's shrine.

In displays and screens, you mesmerize,
Colors vivid, like a painter's sunrise,
A prism of pixels, a visual surprise,
Indium, the artist of digital skies.

In sustainability, you play a key role,
A guardian of Earth, a beacon of soul,

From solar panels to wind turbines that roll,
Indium, the sustainer, making our planet whole.
  Innovation's ally, you take the lead,
In green energy's quest, you plant the seed,
A metal of wonders, in every field,
Indium, the hero, making progress yield.
  Oh Indium, element of grace,
Weaving connections in every space,
Your versatility, a boundless embrace,
Forever cherished, in this cosmic chase.

TWENTY-ONE

ETERNAL FABLE

In the realm of technology, where innovation thrives,
There's a metal that guides and connects our lives.
Indium, the element, so rare and sublime,
With properties that transcend space and time.

A conductor of electricity, it leads the way,
In touchscreens and displays, where its magic holds sway.
With a gentle touch, it responds to our command,
Creating a world where our visions expand.

Indium, the guardian of sustainable dreams,
In solar cells and batteries, its power gleams.
Harnessing the sun's energy, it lights up our days,
In a dance with nature, in renewable ways.

Its shimmering beauty, a feast for the eyes,
As light reflects, creating vibrant skies.

From kaleidoscopic hues to a magnificent gleam,
Indium paints a picture, like a vivid dream.

In the digital realm, it breathes life into screens,
With clarity and sharpness, it captures our dreams.
Indium, the emissary of sustainability's cause,
Reducing energy consumption and ecological flaws.

Oh, Indium, the element of grace and might,
Igniting innovation, casting a radiant light.
A symbol of progress, a conductor of change,
In this technological world, you'll always rearrange.

So let us celebrate Indium's versatile role,
For in its presence, we find beauty and control.
A testament to the wonders of the periodic table,
Indium, forever in our hearts, an eternal fable.

TWENTY-TWO

CONDUCTOR OF CHANGE

In the realm of elements, Indium shines bright,
A conductor of change, a guiding light.
With a lustrous glow, it captures our gaze,
A metal of wonders, in myriad ways.

A catalyst for progress, it leads the way,
Indium, the hero of our modern day.
In every device, in every machine,
Its presence unseen, yet truly serene.

In the world of tech, it weaves a connection,
A bridge between minds, a digital reflection.
From smartphones to laptops, it powers the screen,
Indium, the enigma, behind the unseen.

But its power extends beyond the virtual sphere,
To the realm of sustainability, it steers.

In solar panels, it captures the sun's light,
Indium, the steward, shining so bright.

With each passing day, its influence grows,
A symbol of innovation, as everyone knows.
Indium, the element, so versatile and true,
Guiding humanity to a future anew.

So let us celebrate this metal so rare,
Indium, the beacon, beyond compare.
In its atomic dance, a promise it brings,
Of a greener planet and the joy it sings.

Indium, the element, we honor your name,
For your impact on progress, your role in the game.
In science and industry, you shall remain,
A symbol of hope, a conductor of change.

TWENTY-THREE

BOUND TO BE BRIGHT

In the realm of science, a marvel unfolds,
A shimmering element, Indium it beholds.
A metal so rare, yet it holds great worth,
Guiding us forward, a symbol of rebirth.

Indium, oh Indium, a conductor of light,
In screens and displays, you shine ever so bright.
From smartphones to laptops, you connect us all,
Uniting the world, no matter how big or small.

With a magical touch, you create a bond,
Between glass and circuits, a connection so strong.
Pixels come alive, colors dance and sing,
In the palm of our hands, a technological spring.

But your wonders don't end with screens and displays,
In solar panels, you pave the greenest of ways.

Harnessing the sun's power, you guide the light,
A sustainable future, shining ever so bright.
    Indium, oh Indium, we owe you our praise,
For the innovations you bring, in so many ways.
A catalyst for progress, a bridge to the new,
You sustain our dreams, and make them come true.
    So let us celebrate this element divine,
Indium, oh Indium, a treasure so fine.
With gratitude and awe, we honor your might,
For shaping a world that's bound to be bright.

TWENTY-FOUR

ILLUMINATES US ALL

In the realm of innovation, a metal shines bright,
Indium, the conductor of progress and light.
With atomic number forty-nine in tow,
It sparks a revolution wherever it may go.

From touchscreens to solar cells, it takes the lead,
Indium's versatility, a remarkable deed.
A catalyst for change, a conductor of dreams,
Transforming ordinary materials, it seems.

Indium, the hero, in technology's quest,
Unveiling new horizons, leaving us impressed.
Its soldering prowess, a bond that won't break,
Uniting components, progress it does make.

In the world of sustainability, it plays a vital role,
Indium, the champion of a greener goal.

Efficient and renewable, it paves the way,
Harnessing the sun's power, a brighter day.
    From wind turbines to eco-friendly cars,
Indium's impact reaches near and far.
A conductor of change, an earth-friendly force,
Guiding us towards a sustainable course.
    Oh, Indium, we salute your mighty power,
Your ability to shape a future so much brighter.
With gratitude and awe, we stand tall,
Indium, the element that illuminates us all.

TWENTY-FIVE

BURNING BRIGHT

In screens of vibrant glow, a secret lies untold,
A metal rare and precious, a tale to be unfold.
Indium, the silent hero, in pixels it resides,
Bringing life to images, where innovation hides.

With every touch and swipe, a world is at our hand,
Indium's conductivity, a marvel to understand.
From LCD to OLED, it paves the way for more,
A symphony of colors, it paints on screens galore.

But Indium's worth extends beyond its visual grace,
For sustainability it stands, leaving a greener trace.
In solar panels it finds purpose, harnessing the sun's might,
A renewable energy, to guide us through the night.

Versatile and transformative, Indium takes its role,
From semiconductors to alloys, it plays a vital role.

In progress it is woven, a thread that never fades,
Connecting the components, uniting trades.

So let us celebrate Indium, a metal to admire,
For its soldering capabilities, it sets our hearts on fire.
In screens and solar panels, it shines with endless might,
A testament to progress, a beacon burning bright.

TWENTY-SIX

RISE AND STAND

In the realm of pixels and light,
Where screens come alive, shining bright,
There lies a hidden element, pure and true,
Indium, a marvel, a gift, just for you.

With a lustrous gleam, it takes its place,
Leading us to a digital embrace,
In the touch of a button, a world unfolds,
Indium's magic, a story yet untold.

It dances with electrons, a conductor's delight,
Guiding the current, with precision so right,
In LCD screens, it finds its home,
Transmitting images, where dreams are sown.

But Indium's tale goes beyond the screen,
A symbol of progress, in everything we've seen,

From solar panels, it harnesses the sun,
Powering our world, when day is done.

    A soldering hero, it joins the fray,
Uniting components, in a seamless display,
With strength and flexibility, it paves the way,
For innovation's march, day after day.

    Oh Indium, we marvel at your grace,
Your versatility, a constant embrace,
In every industry, you leave a mark,
A reminder of progress, a spark in the dark.

    So let us give thanks, for Indium's might,
For its role in sustainability, shining so bright,
A beacon of hope, a guide to us all,
Indium, our ally, as we rise and stand tall.

TWENTY-SEVEN

BEAUTIFUL EARTH

In the realm of science, a marvel we find,
A metal so rare, with a radiant mind.
Indium, the element, so versatile and bright,
Igniting innovation, guiding us towards light.

    A conductor of change, it captures the sun,
Transforming its power, a battle half won.
In solar panels, it gleams with grace,
Harvesting energy, a sustainable embrace.

    With soldering might, it binds and connects,
Welding the pieces, ensuring no defects.
In circuitry and screens, its role so grand,
Enabling progress, a future close at hand.

    Indium, oh Indium, a beacon of hope,
Innovating our world, helping us cope.

With its gallium partner, it paves the way,
Inventing the future, every single day.
 From touchscreens to LEDs, its touch is profound,
A catalyst for change, turning the world around.
In labs and factories, its presence is felt,
A reminder of progress, where dreams are dwelt.
 Indium, we thank you for the gifts you bestow,
A metal of wonders, a friend in our show.
With your shimmering essence, you light up our days,
Guiding us forward, in so many ways.
 So let us celebrate, this element divine,
Indium, our ally, in this cosmic design.
With gratitude and awe, we honor your worth,
As the conductor of change, on this beautiful Earth.

TWENTY-EIGHT

SYMBOL OF HOPE

In the realm of elements, a star is born,
Indium, the metal that shines with morn.
A symbol of progress and sustainability,
Harnessing the sun's power with agility.

Indium, the guardian of solar might,
Capturing photons, transforming light.
In solar panels, its purpose is clear,
A catalyst for change, dispelling fear.

With soldering prowess, it binds and connects,
Uniting circuits, technology reflects.
Indium, the conductor, bridging the divide,
Advancing progress with each joint it ties.

From touchscreens to LEDs, its magic unfolds,
Enlightening the world, stories yet untold.

A catalyst in progress, a beacon of light,
Indium guides us through the darkest night.
    In liquid crystals, it reveals the way,
Displaying colors in a vibrant display.
A prism of possibilities, it unveils,
Indium, the alchemist, whose touch never fails.
    Indium, we salute your versatile might,
Transforming industries, igniting our sight.
Grateful we are for your presence so grand,
A symbol of hope, progress at hand.

TWENTY-NINE

ACROSS THE LAND

In the realm of soldering, it's found,
A metal that unites, where bonds resound.
Indium, the element, with grace it shines,
Bringing components together, in soldered lines.

    A conductor of connection, it plays its part,
Uniting the fragments, with a gentle art.
With steady hands, it binds and holds,
Creating pathways, where energy unfolds.

    But beyond the solder, its wonders grow,
In solar panels, it helps the world to glow.
Harnessing the sun's rays, it captures the light,
Indium, the element, shining so bright.

    Versatile and transformative, it takes on new roles,
From touchscreens to semiconductors, it unfolds.

In LCD screens, its magic is displayed,
Indium, the element, a world it has made.
  In every industry, it leaves its mark,
A catalyst for progress, lighting the dark.
From aerospace to medicine, it paves the way,
Indium, the element, leading the day.
  Oh Indium, we sing your praises high,
For the wonders you bring, the dreams you imply.
A conductor of change, a force for good,
Indium, the element, we're grateful you stood.
  In a world that yearns for hope and progress,
You connect and advance, never digress.
Indium, we honor your presence so grand,
A symbol of inspiration, across the land.

THIRTY

WE SHALL SEEK

In the realm of metals, Indium shines bright,
A conductor of change, a beacon of light.
From the depths of the Earth, it does emerge,
A substance so versatile, it's hard to gauge.

    Indium, oh Indium, you connect and unite,
In various industries, you take your flight.
From microchips to LCDs, you play your part,
Guiding our world with your electric art.

    In the world of technology, you stand tall,
With vibrant colors, you captivate us all.
Indium tin oxide, a transparent delight,
Powering displays with a mesmerizing sight.

    But your beauty extends beyond the screen,
In solar panels, you make the world green.

Your conductivity, a gift to progress,
Advancing the world with each new success.
    Indium, oh Indium, we owe you our thanks,
For transforming our lives and breaking the ranks.
A catalyst for change, a metal of might,
You lead us towards a sustainable light.
    So let us raise our voices and sing,
To Indium, the conductor of everything.
For without you, our world would be dull and bleak,
But with your presence, progress we shall seek.

THIRTY-ONE

RADIANT LIGHT

Indium, oh radiant conductor of change,
Innovation's ally, with brilliance untamed.
In your metallic embrace, progress is born,
A testament to the wonders you adorn.

From solar panels, your power unfurls,
Harnessing sunlight, our sustainable pearls.
Through semiconductors, you bring life to machines,
Connecting the world with your shimmering beams.

Touchscreens, with their magic, you effortlessly empower,
Enabling our fingertips to dance and devour.
In LEDs, your luminescence ignites,
Illuminating our nights with celestial lights.

Liquid crystals, a symphony of hues,
Your presence enchants, captivating our views.

Indium, you bridge the realms of science and art,
A catalyst for ingenuity, a masterpiece from the start.
   Oh, how we owe our gratitude to you,
For the wonders you've gifted, so faithful and true.
In every industry, you leave your mark,
A symbol of hope, a beacon in the dark.
   Indium, your name resounds in our hearts,
A testament to progress, where science imparts.
May your legacy endure, forever bright,
Guiding us forward, with your radiant light.

THIRTY-TWO

CONDUCTIVE CHARM

In the depths of the Earth, a treasure lies,
A metal rare, with shimmering guise.
Indium, oh Indium, we sing your praise,
For the wonders you bring, in countless ways.

In solder's embrace, you bind and connect,
Electronic marvels, you help erect.
With precision and skill, you bridge the divide,
Transforming ideas, with each circuit guide.

In the realm of screens, you take center stage,
Displaying colors, with vibrant engage.
From laptops to phones, you illuminate,
Bringing life to pixels, in a symphony of fate.

But your gifts don't end with technology's might,
You aid in medicine, shining so bright.

Radiopaque hero, you guide the way,
In X-ray machines, where shadows sway.
    And in solar panels, you harness the sun,
Transforming its energy, for everyone.
A renewable spark, you help us embrace,
A future sustainable, a brighter place.
    Oh Indium, precious element of grace,
We honor your role, in this cosmic race.
From industry to art, you inspire us all,
With your conductive charm, we stand tall.

THIRTY-THREE

BRILLIANCE

Indium, conductor of progress,
With your noble silvery hue,
You dance on the periodic table,
A catalyst for breakthroughs.
    In the realm of electronics,
You reign supreme, oh Indium,
Your magic touch upon circuitry,
Brings forth a world we can't begin to fathom.
    LCD screens, oh marvel of technology,
You owe your vibrant colors to Indium's grace,
A symphony of pixels, dancing in harmony,
Guiding our eyes through a visual embrace.
    But your prowess doesn't end there,
Oh Indium, you're a gift to solar might,

In photovoltaic cells, you bask in the sun's glare,
Harnessing its energy, shining so bright.
  Conductor of electricity, you are,
A bridge between potentials, a guiding light,
In wires and solder, you conduct with flair,
Bringing power to life, day and night.
  And in the realm of medicine, dear Indium,
You aid us with your X-ray gaze,
Revealing hidden truths, healing within,
Aiding doctors in their caring ways.
  Oh Indium, you hold the key,
To a future that's sustainable and bright,
With renewable energy, we shall be free,
Guided by your brilliance, shining so right.

THIRTY-FOUR

INDIUM'S SIGNIFICANCE

In the realm of elements, there lies a treasure,
A metal rare, with properties to measure.
Indium, the name that shines with grace,
Embodiment of progress in every trace.

Technology's ally, Indium's embrace,
From touchscreens to LEDs, it finds its place.
A conductor of dreams, it paves the way,
Bridging the divides, where progress may sway.

In the world of commerce, it plays its part,
Indium's versatility, a work of art.
Aerospace soars high, with Indium's might,
Medicine thrives, guided by its light.

LCD screens, a vibrant display,
Indium's touch, colors in a grand array.

Catalyst for progress, it leads the pack,
Illuminating paths, never looking back.
Solar panels bask in Indium's glow,
Conducting energy, a sustainable flow.
Advancing a future, where green is the key,
Indium's contribution, a guide for all to see.
Indium, the element of boundless worth,
Connecting the dots, across the Earth.
In technology, medicine, and renewable might,
Indium's significance, a shining light.

THIRTY-FIVE

AWED

In the realm of elements, a gem we find,
Indium, a treasure of a different kind.
A shining star in the periodic chart,
With properties that set it apart.
    In the world of screens, Indium takes its place,
With LCDs, it brings vividness and grace.
Pixels dance, colors bloom with delight,
Indium's conductivity, a shining light.
    Renewable energy, it holds the key,
Solar panels harness its power with glee.
From sun's rays to electricity's might,
Indium's presence, a radiant sight.
    In aerospace, it soars high and free,
Indium alloys, a marvel to see.

From spacecraft to satellites in the sky,
Indium's strength, it never will deny.
  In medicine, a healer it becomes,
Radiopharmaceuticals, its beat drums.
From diagnosis to treatment, it lends a hand,
Indium's touch, a cure that's grand.
  Oh Indium, with gratitude we sing,
For your beauty and versatility, we bring.
A catalyst for progress, a symbol of hope,
In every field, your brilliance we elope.
  Indium, you connect us, you pave the way,
A beacon of light, guiding us every day.
With gratitude in our hearts, we applaud,
Indium, the element, that leaves us awed.

ABOUT THE AUTHOR

Walter the Educator is one of the pseudonyms for Walter Anderson. Formally educated in Chemistry, Business, and Education, he is an educator, an author, a diverse entrepreneur, and he is the son of a disabled war veteran. "Walter the Educator" shares his time between educating and creating. He holds interests and owns several creative projects that entertain, enlighten, enhance, and educate, hoping to inspire and motivate you.

Follow, find new works, and stay up to date
with Walter the Educator™
at WaltertheEducator.com

www.ingramcontent.com/pod-product-compliance
Lightning Source LLC
LaVergne TN
LVHW051959060526
838201LV00059B/3733